# Bonding Systems and Equipment

**Published 2025 by River Publishers**
River Publishers
Broagervej 10, 9260 Gistrup, Denmark
www.riverpublishers.com

**Distributed exclusively by Routledge**
605 Third Avenue, New York, NY 10017, USA
4 Park Square, Milton Park, Abingdon, Oxon OX14 4RN

*Bonding Systems and Equipment* / by Gregory P. Bierals.



Routledge is an imprint of the Taylor & Francis Group, an informa business

ISBN 978-87-7004-816-3 (paperback)

ISBN 978-87-7004-818-7 (online)

ISBN 978-87-7004-817-0 (ebook master)

A Publication in the River Publishers Series in Rapids

# Bonding Systems and Equipment

Gregory P. Bierals
Electrical Design Institute

NEW YORK AND LONDON

# Contents

# Preface

This very important topic of Bonding Systems and Equipment is addressed throughout NEC Article 250, specifically in Part V of this Article. The bonding of equipment, as well as the method of bonding at the service, is covered in detail. In addition, bonding for communications systems, as well as the installation of an intersystem bonding termination device at the service or metering equipment is referenced in 250.94(A),(B). The bonding and grounding requirements for conductive optical fiber cables are specified in 770.100(A),(B).

The bonding and grounding requirements for communications systems, antenna systems, and community antenna television and radio distribution systems are addressed in 800.100, 810.21, and 820.100. The critical equipotential bonding methods for swimming pools are covered in 680.26 and for agricultural building in 547.44(A),(B). And, of course, the equipment grounding and bonding requirements for information technology equipment are addressed in 645.15. The use of insulated equipment grounding conductors and insulated equipment bonding conductors are required in patient care spaces of health care facilities, in accordance with 517.13(B)(1).

These are the subjects that are covered in detail throughout this book in a method that is easy to understand as well as to apply to assure a safe and reliable electrical installation.

As always, I welcome your comments and suggestions.

**Gregory P. Bierals**
**Electrical Design Institute**
**October 1, 2024**

## About the Author

**Gregory P. Bierals** is the president and technical director of Electrical Design Institute. He has presented technical seminar programs on the topics of the National Electrical Code, Designing Overcurrent Protection, NEC Article 240 and Beyond, Grounding Electrical Distribution Systems, Electrical Systems in Hazardous (Classified) Locations, and Grounding and Bonding Solar Photovoltaic and Energy Storage Systems. These courses have been offered through the auspices of the University of Wisconsin, George Washington University, the University of Alabama, the University of Toledo, Appalachian State University and North Carolina State University. He has also designed and supervised the installation of electrical systems and equipment in Health Care Facilities throughout Africa and North Korea. His published books include The NEC and You, Perfect Together, Grounding Electrical Distribution Systems, Designing Overcurrent Protection, NEC Article 240 and Beyond. and Grounding and Bonding Photovoltaic and Energy Storage Systems. He has recently started a short books series for River Publishers.

# CHAPTER 1

## Bonding Systems and Equipment

This is the third book in our short book series and the topic is, "Bonding Systems and Equipment".

But, before we begin, let us examine 250.110, which covers the grounding requirements of "fixed equipment", that is, "equipment that is fastened or otherwise secured at a specific location" (Article 100). Normally, the exposed, non-current carrying metal parts of fixed equipment that are supplied by, or including, conductors or components that are likely to become energized, are required to be connected to an equipment grounding conductor under the following conditions:

If within 8 feet (2.5 m) vertically or 5 feet (1.5 m) horizontally of ground or grounded metal objects and subject to contact by persons.

However, where the fixed equipment is installed above this height (8 feet vertically (2.5 m) or more and 5 feet (1.5 m) horizontally from ground or grounded metal objects, and not subject to contact by persons, this fixed equipment may not be required to be grounded (250.110(1)). And, where this equipment is installed in this manner and the equipment is subject to contact by people, *bonding* the equipment to reduce the effects of voltage differences between this equipment is a requirement of 250.92(A) and 250.96(A). So, this equipment may not be grounded due to its physical isolation, but *bonding* the equipment may still be a requirement.

As we begin our study of this important topic, an analysis of the definitions of the terms that are directly associated with this subject from Article 100 are identified and explained here.

- **Bonded (bonding):** Connected to establish electrical continuity and conductivity.
- **Bonding conductor (bonding jumper):** A conductor that ensures the required electrical conductivity between metal parts that are required to be electrically connected.
- **Bonding Jumper, Equipment:** The connection between two or more portions of the equipment grounding conductor. The equipment bonding jumper may be used as a means of connecting the metal raceways that contain the conductors of a paralleled feeder. Based on the ampere rating of the feeder overcurrent device, and a reference to Table 250.122, the *minimum* size of the equipment bonding jumper may be determined. However, the actual size of this bonding jumper can only be determined by calculating the available fault current and determining the operating characteristics of the overcurrent device (clearing time and let-through current), as well as the melting or fusing current of the equipment bonding jumper.
- **Bonding Jumper, Main:** The connection between the grounded circuit conductor and the equipment grounding conductor, or the supply-side bonding jumper, or both at the service. The *minimum* size of this bonding jumper is based on Table 250.102(C)(1). But, once again, the actual size of the main bonding jumper is determined by the fault current (the current delivered at a point on the system during a short-circuit condition) that the main bonding jumper will carry until the downstream faulted circuit overcurrent device or the service overcurrent device opens to clear the downstream short-circuit or ground-fault. For a system that is solidly grounded, probably 90% of the fault current will return to the source (transformer, generator, etc.) through the grounded service conductor and the remaining 10% through other conducting paths, including the path through the grounding electrode conductor, through the earth to the grounding electrode at the source, and then through the grounded conductor to the system neutral point or the grounded phase of a corner-grounded delta system.

In order to limit the voltage-rise about the earth potential on the electrical distribution system, including the grounding system, the length of the grounding electrode conductor and the method of installation of this conductor is an important consideration.

The length of this conductor from the service equipment to the grounding electrode must be limited and installed as straight as practicable, avoiding any unnecessary and especially sharp bends. Where bends in this conductor are necessary, the bending radius of the grounding electrode conductor must be no less than 8 inches (203 mm). This is due to higher frequency lightning currents (3 kHz–10 mHz) and the "skin effect" of the alternating currents at this high frequency (NFPA 780).

- **Bonding Jumper, Supply-side:** A conductor installed on the supply side of a service or within a service equipment enclosure(s), or for a separately derived system, that ensures the required electrical conductivity between metal parts that are required to be electrically connected. The *minimum* size of this bonding jumper is identified in Table 250.102(C)(1).

There is a concern where the supply-side bonding jumper is on the supply side of the service. The only overcurrent protection for this bonding jumper and the ungrounded service conductors may be the overcurrent protection on the primary side of the supply transformer. And, if this transformer is a three-phase delta-to-wye connected system, where there is a 30 degree phase-shift from primary-to-secondary, and the primary voltage leads the secondary voltage by 30 degrees, the only overcurrent protection for the secondary conductors, including the supply-side bonding jumper may be the primary overcurrent device. This overcurrent device may be sized at 3–6 times (or more) of the full-load transformer primary current. Section 230.82(1) permits "cable limiters" to be installed on the supply-side of the service disconnect and these devices are current-limiting. Which means that they will operate to clear a fault in less than half a cycle (0.008 seconds) when interrupting fault currents in their current interrupting range (short-circuit currents reach a peak during the first one-half cycle of the fault). So, the cable limiters may clear the short-circuit before it reaches its peak. This type of overcurrent protection will isolate a faulted cable where the service conductors are installed in parallel so that the service supply system will remain continuous. In addition, the cable limiters can be used to provide the proper overcurrent protection for the service equipment in accordance with the short-circuit withstand rating of this equipment (110.24(A), 110.10).

The supply-side bonding jumper is often related to separately derived systems (250.30(A)(2)). Where the source of the separately derived system and the first disconnect are in separate enclosures, there will be a supply-side bonding jumper installed from the source to the first disconnecting means. This conductor will normally be of a smaller size than the ungrounded conductors, as its *minimum* size is determined in accordance with 250.102(C)(1).

If a *system bonding jumper* is installed at the source of the separately derived system, the connection of the grounding electrode conductor to the grounding electrode (system) will be made at the source. If the system bonding jumper is installed with the feeder conductors to the first system disconnect or overcurrent device, the connection of the grounding electrode conductor to the grounding electrode (system) is made at this location. This provision is meant as a means of limiting the length of the grounding electrode conductor, in that, if the grounding electrode is closer to the source than it is from the first disconnecting means, then the connection to the grounding electrode should be made at this location. If the grounding electrode is closer to the first disconnecting means than it is from the source, the connection should be made at this location. In either case, we stress that the grounding electrode conductor should be installed as short and straight as practicable, avoiding unnecessary and especially sharp bends (250.4(A), Informational Note No. 1).

The grounding electrode for the separately derived system will be the building or structure grounding electrode system in accordance with (250.30(A)(4)). This will satisfy 250.50 and 250.58, as the grounding electrode for the separately derived system will be bonded to the other grounding electrode system within the same building or structure.

- **Bonding Jumper, System:** The connection between the grounded circuit conductor, and the supply-side bonding jumper, or the equipment grounding conductor, or both, at a separately derived system (250.30(A)(1)). As we have stated for the supply-side bonding jumper, this connection is made at the source or at the first system disconnecting means that is supplied from the source. However, if the source of the separately derived system is outdoors, the system bonding connection and grounding electrode conductor connection to the grounding electrode must be made at the source (250.30(C), 250.50, 250.52).

Outdoor locations are more susceptible to lightning induced influences and, therefore, the earth connection should be made outdoors as well, with connections that are short and straight, and through the use of the connection methods specified in 250.8(A).

In any event, the main bonding jumper and system bonding jumper are installed in accordance with 250.28 and may be of copper, aluminum, or copper-clad aluminum and in the form of a wire, bus, or screw (listed and indentified with a green finish that is visible when installed (110.3(B), 110.3(C), 110.10).

We begin our analysis with the terms and definitions associated with bonding systems and equipment. The word "bond" as defined in the Oxford Dictionary, is "a connection between surfaces or objects that have been joined together, especially by means of an adhesive substance, heat, or pressure". The joining of electrical conductors by heat (solder) was a common practice in the electrical industry many years ago. I still remember some of the older electricians and their alcohol torches that they had used with solder to splice conductors. Alloys that are commonly used for electrical soldering have melting temperatures that range from 183–188 °C (361–370 F).

Short-circuit or ground-fault currents may cause an increase in conductor temperatures that exceed these melting temperatures. So, connections that rely on solder are not acceptable (250.8(B)).

The joining of conductors or the connection of conductors to steel by the exothermic welding system were developed by Professor Charles Cadwell in 1938–1939 at the Case School of Applied Science (Case Western Reserve University) in Cleveland, Ohio. This welding process was initially invented by Hans Goldschmidt in 1895. Professor Cadwell further developed this process with copper (the reduction of copper oxide by aluminum). And, its original use was to connect bonding jumpers to the rails of railroad tracks. The use of this process expanded over time to include the exothermic welding of electrical conductors and conductors to steel. This process has been known as Cadweld, in recognition of Professor Cadwell (Erico Corporation, now Pentair Plc). The name Erico stands for "Electric Railway Improvement Company", dating to 1903. Of course, there are many examples of electrical connections that rely on some form of pressure and/or heat (110.14, 110.14 (A), 110.14(B)).

The term "bonding" from Article 100, states "connected to establish electrical continuity and conductivity". Also, from 250.4(A)(3), "normally non-current carrying conductive materials enclosing electrical conductors and equipment shall be connected together and to the electrical supply source in a manner that establishes an effective ground-fault current path". According to 250.4(A)(5), this conducting path must create a low-impedance circuit, and, in so doing, causing the operation of the circuit overcurrent device in such a way that the circuit components are protected from the extensive damage associated with the short-circuit or ground-fault. For ungrounded or impedance grounded systems, ground detectors will identify the ground-fault and the maintenance staff can initiate an orderly shutdown of the affected electrical system (110.10, 250.20(A), 250.20(B), 250.21(B)).

From 250.4(A)(5), the effective ground-fault current path is defined as an electrically conductive path of low impedance that will facilitate the operation of the overcurrent device for solidly grounded systems or the ground detector for ungrounded or impedance grounded systems. This low impedance path must be capable of safely carrying the maximum ground-fault current likely to be imposed on it from any point where a ground-fault occurs to the system source.

This conducting path may be through one of the metal raceways identified in Chapter 3, or a *properly sized* equipment grounding conductor (250.122, minimum size), or both.

Certainly, the ability of this effective ground-fault current path to perform these functions must be determined by the actual circuit conditions and not merely by selecting the conductor size from Table 250.122.

This process includes calculating the ground-fault current, determining the operating characteristics of the overcurrent device, and providing the proper overcurrent device that will assure compliance with 110.10, that is, to permit the overcurrent device to clear the fault without the occurrence of extensive damage to the circuit components on a system that is solidly grounded. For ungrounded or impedance grounded systems, the path for fault current must be a low impedance path that will facilitate the operation of the overcurrent devices in the event that a second fault occurs on another phase before the first fault is cleared.

250.90 states that bonding shall be provided if necessary to ensure electrical continuity and the capacity to conduct safely any fault current likely to be imposed.

This statement is almost exactly how the term "bond" was once defined in Article 100.

So, we know from the term "bonded" (bonding) from Article 100 and the term "bonding" from 250.90, that the bonding conductor connection ensures the continuity of the conducting path, but also there is the necessity that this conducting path will be capable of conducting the imposed fault current without damage.

**Example**

So, let us assume that we have a single-phase, dry-type transformer, 50 kVA, 240–120 volts, and the transformer marked impedance is 1.50% (Z).

In this example, we have selected 4/0 AWG-THHN-copper conductors from Table 310.16 (75 °C–230 amperes), including the neutral conductor. The terminal connections are listed at 75°C (110.14(A), 110.14(C)(1)(b)(1)).

UL 1561-Transformers 25 kVA and larger have a + or − 10% impedance tolerance.

So, based on worst case conditions, multiply the percent impedance by 0.90 to calculate the worst case fault-current.

$$\begin{array}{r} 1.5 \\ \times 0.90 \\ \hline 1.35 \end{array} \qquad \frac{100}{1.35} = 74.07$$

$$\frac{50 \times 1000}{240 \text{ volts}} = 208.33 \text{ amperes}$$

$$\begin{array}{rl} 208 & \text{amperes} \\ \times 74.07 & \\ \hline 15,431.25 & \text{amperes} \end{array}$$

This is the available fault current at the transformer secondary terminals (line-to-line)

$$\begin{array}{rl} 15,431.25 & \text{amperes} \\ \times 1.5 & \\ \hline 23,146.88 & \text{amperes} \end{array}$$

This is the available fault current (line-to-neutral) at the transformer secondary terminals.

Note: The fault current, line-to-neutral, is higher than the line-to-line fault current at the single-phase center-tapped transformer, where the secondary conductors, including the neutral conductor, are the same size.

There will be a terminal bar for the connection of all bonding and grounding connections and this terminal bar is secured inside the transformer enclosure, but not blocking any ventilating openings (450.10(A)). This terminal bar for bonding and grounding conductor terminations is required by 450.10, because the metallic transformer enclosure is not normally identified and listed for grounding and bonding connections (250.8(A)(2)).

The terminal bar within the transformer enclosure will have a connection from the equipment grounding conductor that is run with the transformer primary conductors. If the primary conductors are installed in a metallic raceway and this raceway is the primary feeder equipment grounding conductor (250.118(A)), there will be a connection from this metal raceway to the transformer terminal bar. And, its *minimum size* is in accordance with 250.122.

The size of the supply-side bonding jumper has been selected from Table 250.102(C)(1) and, in this case, it is 2AWG copper.

A flexible metal conduit extends from the transformer enclosure to the metal enclosure of a 225 ampere, 2-pole molded-case circuit breaker.

The size of the flexible metal conduit for 3-4/0AWG-THHN insulated conductors and a 2AWG-THHN insulated supply-side bonding jumper is as follows:

**Table 5 – Chapter 9**

4/0AWG-THHN conductors: $0.3237 \text{ sq. in.} \times 3 \; = \; 0.9711 \text{ sq. in.}$

2AWG-THHN conductor: $0.1158 \text{ sq. in.} \times 1 \quad = \underline{0.1158 \text{ sq.in.}}$

$1.0869 \text{ sq.in.}$

The flexible metal conduit is not an equipment grounding conductor in accordance with 250.118(A)(5)(b), so there will be a supply-side bonding jumper installed with the 4/0 THHN copper secondary conductors to the 225 ampere circuit breaker enclosure. This bonding jumper completes the connection from the grounded (neutral) conductor to the equipment ground terminal bar in the 225 ampere circuit breaker enclosure.

**Table 4 – Chapter 9**

2 inch Flexible Metal Conduit (Article 348).

**Note:** The supply-side bonding jumper is insulated to prevent arcing within the flexible metal conduit as there will be a difference in potential between this bonding jumper and the flexible metal conduit during a ground-fault, and the possible internal arcing between this bonding jumper and the flexible metal conduit will damage the insulation on the other conductors within this conduit.

The supply-side bonding jumper and the system bonding jumper are connected to the grounding terminal bar within the transformer enclosure.

A grounding electrode conductor connects this transformer terminal bar to the grounding electrode (system). The grounding electrode for the separately derived system will be the same grounding electrode (system) that is used for the service equipment, communications equipment and other systems within the same building or structure (250.30(A)(4), 250.50, 250.58).

In addition, the interior metal water piping system in the area that is served by the separately derived system is bonded to the grounded conductor of the separately derived system. This bonding connection is made at the nearest accessible point of the metal water piping system. The bonding conductor is sized in accordance with Table 250.102(C)(1), based on the cross-sectional area of the ungrounded conductors supplied from the transformer secondary.

In our example, the grounding electrode for the separately derived system is the metal in-ground support structure that is used for the service equipment and other systems within this building. So, there is no need for an additional bonding conductor from the transformer terminal bar to the building structural metal (250.104(D)(2), Exception No. 2).

The size of the grounding electrode conductor from the transformer terminal bar to the building structural metal is in accordance with 250.30(A)(5) and 250.66, and, in this case, it will be 2AWG copper. This conductor will be installed in accordance with 250.4(A)(1). That is, it is run as short and straight as practicable, avoiding unnecessary and especially sharp bends. In consideration of NFPA 780, any bend in this conductor should include a radius of not less than 8 inches (203 mm), and the number of bends should be limited in recognition of the high frequency of lightning currents (3 kHz–10 mHz).

There will be a *system bonding jumper* connected from the transformer secondary neutral point to the grounding terminal bar within the transformer enclosure.

The system bonding jumper size is in accordance with 250.102(C)(1), based on the cross-sectional area of the transformer secondary conductors.

In this example, based on the transformer full-load secondary current, we are using 4/0 THHN, copper conductors, 260 amperes, 90 °C–230 amperes–75 °C, as the transformer secondary terminals have a 75 °C temperature limit (110.14(C)(1)(b)(1)).

Based on Table 250.102(C)(1), the system bonding jumper size is 2AWG copper. There is a 225 ampere, molded-case circuit breaker at the transformer secondary and the 4/0AWG (107.22 $mm^2$) copper secondary conductors extend for a length of 4 feet (1.2 m) in a flexible metal conduit to a 2-pole, 225 ampere molded-case circuit breaker. This circuit breaker enclosure requires at least the minimum wire-bending space at the circuit breaker terminals in accordance with Table 312.6(B)(2). For the 4/0AWG conductor size, the wire bending space for the line and load connections will be 7 inches (178 mm) from the line and load terminals to the enclosure wall.

The minimum size of the supply-side bonding jumper is taken from Table 250.102(C)(1). In this case, 2AWG copper.

The flexible metal conduit is terminated in listed fittings (250.118(A)(1)(5)(a), 110.3(B), 110.3(C)).

The 2AWG supply-side bonding jumper is terminated on the required terminal bar that is secured inside the transformer enclosure and bonded to the transformer enclosure (450.10(A), 240.12).

With the available fault-current at the transformer secondary terminals calculated at 23,146.88 amperes, we will check the fusing current of the 2AWG copper system bonding jumper and the supply-side bonding jumper, based on the fault clearing time of the 225 ampere molded-case circuit breaker at the transformer secondary, which has been determined to be 0.025 seconds.

2AWG copper-66,360 circular mils

$$\frac{66,360 \text{ cm}}{16.19} = 4099 \text{ amperes} - 5 \text{ seconds}$$

**Note:** One ampere for every 16.19 circular mils of the conductor cross-sectional area for 5 seconds.

$$4099 \text{ amperes } \; 4099 \text{ amperes } \; 5 \text{ seconds}$$

$$= 84,009,005 \text{ ampere} - \text{squared seconds}$$

$$\frac{84,009,005}{0.025 \text{ seconds}} = 3,360,360,200$$

$$\sqrt{3,360,360,200} = 57,968.61 \text{ amperes}$$

So, the fusing (melting) current of the 2AWG copper system bonding jumper and the supply-side bonding jumper for 0.025 seconds is 57,968.61 amperes, and the line-to-neutral fault current has been determined to be 23,146.88 amperes.

**Table 1.1:** Fusing or melting current

| | | | | 1083° C. MAXIMUM | | | |
|---|---|---|---|---|---|---|---|
| AWG | NORMAL | 5 second | 1 second | 1 cycle – 0.016 sec. | 1/2 cycle – 0.008 sec. | 1/4 cycle – 0.004 sec. | 0.002 sec. 0.002 sec. |
| | 75 °C | 1083 °C | 1083 °C | 1083 °C | 1083 °C | 1083 °C | 1083 °C |
| 14 | 20 A | 254 A | 568 A | 4,490 A | 6,3 50 A | 8,980 A | 12,700 A |
| 12 | 25 A | 403 A | 901 A | 7,124 A | 10,075 A | 14,248 A | 20,150 A |
| 10 | 35 A | 641 A | 1433 A | 11,331 A | 16,025 A | 22,663 A | 32,050 A |
| 8 | 50 A | 1020 A | 2281 A | 18,031 A | 25,500 A | 36,062 A | 51,000 A |
| 6 | 65 A | 1621 A | 3625 A | 28,656 A | 40,525 A | 57,311 A | 81,050 A |
| 4 | 85 A | 2578 A | 5765 A | 45,573 A | 64,450 A | 91,146 A | 128,900 A |
| 3 | 100 A | 3312 A | 7406 A | 58,548 A | 82,800 A | 117,097 A | 165,000 A |
| 2 | 115 A | 4101 A | 9170 A | 72,461 A | 102,475 A | 144,922 A | 204,950 A |
| 1 | 130 A | 5169 A | 11,558 A | 91,376 A | 129,225 A | 183,105 A | 258,950 A |
| 1/0 | 150 A | 6523 A | 14,586 A | 115,311 A | 163,075 A | 230,623 A | 326,150 A |
| 2/0 | 175 A | 8221 A | 18,383 A | 145,328 A | 205,525 A | 290,656 A | 411,050 A |
| 3/0 | 200 A | 10,364 A | 23,175 A | 183,2U A | 259,100 A | 366,423 A | 518,200 A |
| 4/0 | 230 A | 13,070 A | 29,225 A | 231,047 A | 326,750 A | 462,094 A | 653,500 A |
| 250 kcmil | 255 A | 15,442 A | 34,529 A | 272,979 A | 386,050 A | 545,957 A | 772,100 A |
| 300 kcmi | 285 A | 18,530 A | 41,434 A | 327,567 A | 463,250 A | 655,134 A | 925,500 A |
| 350 kcmi | 310 A | 21,618 A | 48,339 A | 382,156 A | 540,450 A | 764,312 A | 1,080,900 A |
| 400 kcmi | 335 A | 24,707 A | 55,247 A | 436,762 A | 617,675 A | 873,524 A | 1,235,350 A |
| 500 kcmil | 380 A | 30,883 A | 69,056 A | 545,939 A | 772,075 A | 1,091,879 A | 1,544,150 A |

This point-to-point fault current calculation is used to determine the validity of the system bonding jumper and the supply-side bonding jumper for the 50 kVA–240/120 volt transformer (250.30(A)(1), 250.30(A)(2)), 2AWG copper, Table 250.102(C)(1)) with an identified percent impedance of 1.5% and a worst case percent impedance of 1.35%, based on UL 1561 (−10%), (110.10).

This same procedure is used to determine the short-time (fusing current) current-carrying capacity of the Main Bonding Jumper within the service equipment (250.24(C), 250.28(A),(B),(C), 250.102(C)(1)), with the minimum size in accordance with Table 250.102(C)(1), based on the cross-sectional area of the ungrounded service conductors. And, the actual size based on the available fault current at the service equipment and the duration of this fault current in accordance with the clearing time of the service overcurrent device(s) and the fusing or melting current of the main bonding jumper (Table 1). The main bonding jumper must be able to *safely* carry the fault current until the appropriate overcurrent device operates to clear the fault (110.10).

The secondary of the 50 kVA transformer is a separately derived system. This system has overcurrent protection. In this example, this overcurrent device is a 225 ampere, 2-pole, molded-case circuit breaker that is supplied from the transformer secondary (250.32(B)(2)). A System Bonding Jumper has been installed in accordance with 250.30(A)(1), 250.102(C)(1), in this case, at the source of the separately derived system.

There must be a secondary overcurrent device for this transformer because of 240.4(F). This transformer has a 3-wire secondary and the primary overcurrent device cannot be used to protect the secondary conductors due to the possible and likely unbalanced load supplied by the transformer secondary. Because the primary overcurrent device may have an ampere rating of 250% of the transformer primary full-load current rating (Table 450.3(B)), and if the primary voltage is 480 volts, this would be:

$$\frac{50,000\ \text{VA}}{480\ \text{V}} = 104\ \text{A} \qquad \begin{array}{r} 104\ \text{A} \\ \times 2.50 \\ \hline 260\ \text{A} \end{array}$$

Note: The use of the next larger overcurrent device is not permitted where primary and secondary overcurrent protection is provided.

So, the transformer primary overcurrent device may have an ampere rating of 250 amperes. Using this ampere rating will also determine the ampere rating of the transformer primary conductors (450.3, Informational Note No. 1 and 240.4). The use of a lower rated overcurrent device, where possible, will permit the use of smaller conductors and better overcurrent protection for the transformer.

It should also be noted that 240.4(F) applies to 3-phase transformers, where the secondary is 3-phase, 4-wire, wye connected and the primary is delta connected. Due to the 30 degree phase shift from primary to secondary, the secondary conductors require properly sized overcurrent protection. And, the need for overcurrent protection for the secondary conductors also applies to a 3-phase, 4-wire, delta system.

Finally, where the transformer secondary overcurrent device is required, it should be installed close to the transformer secondary terminals in recognition that the conductors on the line-side of the secondary overcurrent device are only provided overcurrent protection by the primary overcurrent protective device, which may be located at the point where the primary conductors receive their supply and not at the location of the transformer.

Table 450.3(B) permits the secondary overcurrent device to be rated at 260 (250) amperes. However, we are using a 225 ampere molded-case circuit breaker in this example.

The available fault-current, line-to-line, at the line terminals of the 225 amperes circuit breaker, is 14,656 amperes.

The available fault-current, line-to-neutral at the 225 ampere circuit breaker is 20,041 amperes, and this establishes the interrupting rating of the 225 ampere circuit breaker, in accordance with 110.9.

The length of the secondary conductors are 6 feet. The fault-current at the line terminals of the 225 ampere circuit breaker is as follows:

$$\frac{2 \times 6 \text{ feet} \times 15,431.25 \text{ A}}{15,082 \text{ ("C") value} \times 1 \times 240 \text{ V}} = \frac{185,175}{3,619,680} = 0.0512$$

The “C” values for conductors are equal to one over the impedance per foot of the conductor, as per the resistance and impedance values in IEEE Standard 241 (Gray Book).

$$\text{Multiplier} = \frac{1}{1 + 0.0512} = 0.9513$$

$$\begin{array}{r} 15,431.25 \text{ A} \\ \times\ 0.9513 \\ \hline 14,679.75 \text{ A} \end{array} \quad - L - L$$

$$\frac{2 \times 6 \text{ feet} \times 23,146.88 \text{ A}}{15,082 \times 1 \times 120 \text{ V}} = \frac{277,762.56}{1,809,840} = 0.1535$$

$$\text{Multiplier} = \frac{1}{1 + 0.1535} = 0.8669$$

$$\begin{array}{r} 23,146.88\ A \\ \times 0.8669 \\ \hline 20,066\ A \end{array} \quad -L-N$$

A feeder extends for a length of 75 feet (23 m) from the 225 ampere, 2-pole circuit breaker to an enclosed panelboard, consisting of 3-4/0AWG-THHN insulated copper conductors in a 2 inch (metric designator 53, Table C.4) rigid metal conduit which serves as the equipment grounding conductor (250.118(A)(2)).

The length of the metal raceway and its cross-sectional area that serves as the transformer secondary equipment grounding conductor must be determined and not merely selected from 250.118(A) (see Georgia Institute of Technology Information on the lengths of RMC, IMC, and EMT and various sizes of these raceways where the fault current is 5 times the ampere rating of the circuit overcurrent device.

The available fault current at the panelboard is as follows:-

$$\frac{2 \times 75 \text{ feet} \times 14,679.75 \text{ A}}{15,082 \times 1 \times 240 \text{ V}} = \frac{2,201,962.50}{3,619,680} = 0.6083$$

$$\text{Multiplier} = \frac{1}{1 + 0.6083} = 0.6217$$

$$\begin{array}{r} 14,679.75 \text{ A} \\ \times 0.6217 \\ \hline 9126.40 \end{array} \quad \text{amperes} - L - L$$

$$\frac{2 \times 75 \text{ feet} \times 23,146.88 \text{ A}}{15,082 \times 1 \times 120 \text{ V}} = \frac{3,472,032}{1,809,840} = 1.9184$$

$$\text{Multiplier} = \frac{1}{1 + 1.9184} = 0.3426$$

$$\begin{array}{r} 23,146.88 \text{ A} \\ \times 0.3426 \\ \hline 7930.12 \end{array} \quad \text{amperes} - L - N$$

The available fault current (line-to-line) is 9126.40 amperes and 7930.12 amperes line-to-neutral at the panelboard. This panelboard has a 225 ampere, molded-case circuit breaker as its main protection, even though 408.36 does not require this protection, as there is a 225 ampere circuit breaker at the transformer secondary. The interrupting rating of this circuit breaker must be at least 9126.40 amperes (10,000 amperes) to satisfy 110.9.

The interrupting rating of the individual branch circuit and feeder overcurrent devices in this panelboard should also be 10,000 amperes.

Section 408.38 states that where the available fault current exceeds 10,000 amperes (only 9126.40 amperes in this example), the panelboard and enclosure combination shall be evaluated for the application by the authority having jurisdiction.

Section 408.6 requires that this panelboard must have a short-circuit current rating not less than the available fault current (9126.40A).

## Communications Systems

The bonding of the communications systems, as well as the grounding of this system is addressed in 800.100. The bonding or grounding electrode conductor must be listed and may be insulated, covered, or bare (800.100(A)(1)). This conductor is normally copper or another corrosion-resistant material. It may be solid or stranded (800.100(A)(2)). The bonding or grounding electrode conductor is of critical concern and 800.100(A)(4) limits the length to 20 feet (6.0 m) for one and two family dwelling units. Of course, limiting the length of these conductors is in recognition of the high frequency lightning current that may flow through the bonding conductor between separate grounding electrode systems or through the grounding electrode conductor to the grounding electrode (system). The frequency of lightning currents range from 3 kHz to 10 mHz and the duration of the lightning current is from 2–10 microseconds. The high ac frequency of this lightning current produces an increase in the skin effect of this current flow. This increases the impedance of the conducting path and is the primary reason for limiting the length of the bonding or grounding electrode conductor to no more than 20 feet (6.0 m). This will reduce the voltage difference between the communications system and other connected systems, as well as between these systems and the earth. In fact, the bonding and grounding connections would be more effective if the conductors were flat, as these conductors will expose more of their cross-sectional area to their surface as compared to conductors that are round. This is the reason that a signal reference grid is constructed of flat copper conductors as a means of reducing voltage differences between equipment in an Information Technology Equipment room. For example, a 4/0AWG (107.22 $mm^2$) copper conductor has a dc resistance of 0.0608 ohms per 1000 feet of length (uncoated copper at 75 °C) and an ac impedance of 0.80 ohms to neutral at 0.85 power factor (1000 feet) in a steel conduit (Tables 8 and 9, Chapter 9). However, at 10 mHz, and only 10 feet long, the ac impedance of the 4/0AWG copper cable is *232 ohms*. This conductor cannot equalize the potential difference across its two ends. Section

800.100(A)(5) requires the bonding or grounding electrode conductor to be run in as straight a line as practicable. This provision relates to 250.4(A)(1) and Informational Note No. 1 and Informational Note No. 2. That is, the routing of the grounding electrode and bonding conductors are so that they are not any longer than necessary to complete the connection. Bends and loops, especially sharp bends in these conductors, will increase the impedance of the bonding and grounding electrode conductors due to skin effect and unnecessary length. Section 800.100(D) requires the bonding connection from the grounding electrode for the communications system and the power grounding electrode system to be no smaller than 6AWG copper or equivalent. Informational Note No. 2 states that the bonding of the grounding electrodes of different system will limit potential differences between these systems and their associated wiring systems.

An incident that occurred in the State of New Jersey many years ago involved a dwelling occupancy that had a land-line telephone system. The telephone system was connected to a grounding electrode (ground rod, 800.100(A)(1),(2),(3)). The service entrance system had the grounded service conductor connected to the grounding electrode system in accordance with 250.24(A)(1),(2). The problem was that there was no bonding conductor connection between the two grounding electrodes (250.50, 250.58, 800.100(A)(1),(2),(3),(4),(5),(6)). There was thunderstorm activity in this area at this time and a young man was in his bedroom. This bedroom had a telephone and other electrical equipment (television, lamps, etc.). Later that evening his parents found him deceased. Either the telephone system or the electrical service was subjected to a significant voltage rise due to lightning, and this young man came in contact with both systems at this time. The voltage difference may have been on the order of several hundred thousand volts (or more) and this voltage difference caused his immediate death. Incidents such as these have occurred many times because of significant voltage differences.

If the bonding conductor(s) and grounding electrode conductors are exposed to physical damage, they must be protected in accordance with 800.100(A)(6), 250.64(E)(1). If these conductors are in ferrous metal raceways, both ends of this raceway must be bonded to the internal conductor in order to reduce the impedance of the connection to ground.

Not only are the grounding electrodes of different systems required to be properly bonded. The bonding must be done in such a way as to limit the voltage difference between the separate grounding systems. For example the bonding and grounding electrode conductors must be properly sized, as well as installed as short and straight as possible, eliminating any unnecessary bends and loops. NFPA 780 (2020), Standard for the Installation of Lightning Protection Systems, requires the bends of the main conductor and down conductors to have a radius of no less than 8 inches (203 mm).

Sections 250.50, 250.58, 250.60, 250.70, 250.92, 250.94, 800.100, 810.21, and 820.10 require the effective bonding of separate grounding electrodes.

## NFPA 780: Standard for the Installation of Lightning Protection Systems

Building and structures that are up to 75 feet (23 m) in height use Class I materials, and the conductors are 57,400 circular mils in size. For building or structures of over 75 feet (23 m) in height, Class II materials are used, and these conductors are 115,000 circular mils in size.

57,400 circular mils (equivalent size, 2AWG, 66,360 circular mils)

115,000 circular mils (equivalent size, 2/0AWG, 133,100 circular mils)

It should also be noted that the main conductor that is used to bond the strike termination devices on a building or structure and the lightning down conductors that are used to *bond* the strike termination devices to the grounding electrode system (ground ring) are not constructed of normal building wire. These conductors are stranded in a manner that resembles a weave pattern in recognition of the skin effect of the high frequency lightning current. If the conductor that is used to bond the grounding electrodes of the grounding electrode system is this type of conductor, the voltage differences between the grounding electrodes will be reduced as compared to the used of normal building wire.

The bonding for communications systems is addressed in 250.94. An "Intersystem Bonding Termination Device" is required as a means of connecting the bonding conductors of different systems. This device is required to be external to enclosures at the service or metering equipment enclosure and at the disconnecting means for buildings or structures that are supplied by a feeder or branch circuit. The IBT is required to be accessible, it must have no less than three intersystem bonding conductor terminations, and it must be installed in such a way as to not block the opening of the enclosure for the service equipment, feeder or branch circuit equipment, or metering equipment (250.94(A)).

The intersystem bonding termination device at the service equipment is required to be securely mounted to the service equipment enclosure, to a metal meter enclosure, to a nonflexible metal service raceway, or to the metal enclosure for the grounding electrode conductor (250.64(E)(1)), with a 6AWG or larger copper conductor.

Where the building or structure is supplied by a feeder or branch circuit, the intersystem bonding termination device must be electrically connected to the metal enclosure for the building or structure disconnecting means, or connected to the metal enclosure for the grounding electrode conductor (250.64(E)(1)) with a 6AWG or larger copper conductor.

Equipment that is listed as bonding or grounding equipment may be used for this purpose (250.94(A)(5)).

In an existing building or structure, an Intersystem Bonding Termination Device is not required. However an external means to connect bonding or grounding electrode conductors is permitted at the service, feeder, or branch circuit disconnecting means. This external bonding means may consist of:

(1) Exposed nonflexible metal raceways
(2) An exposed grounding electrode conductor
(3) An approved means for an external connection of a copper or other corrosion-resistant bonding or grounding electrode conductor to the grounded raceway or equipment (250.94(A), Exception).

We have stressed the importance of intersystem bonding because of the voltage differences between systems, especially due to high frequency lightning currents. (770.100, Conductive Optical Fiber Cables, 800.100, Communications Systems, 810.21, Radio and Television Equipment, and 820.100 Network-Powered Broadband Communications Systems). With the use of the proper bonding and grounding methods between different systems, not only is the overall resistance-to-ground reduced, due to the parallel arrangement of the grounding electrode system, but an added benefit is the reduction of electromagnetic interference (noise) (EMI) on the communications system.

Section 250.96 requires the metal raceways, cable trays, cable armor, cable sheath, enclosures, frames, fittings, and other metal non-current carrying parts that are used as equipment grounding conductors to be properly bonded to ensure electrical continuity and their ability to safely conduct any fault current that may be imposed. Once again, the fact that the metallic raceway or cable assembly is referenced in 250.118(A) as a type of equipment grounding conductor does not mean that these conductors provide the effective ground-fault current path required by 250.4(A)(5) and 250.4(B)(4).

The ground-fault current must be calculated, the impedance of the equipment grounding system must be determined, and the operating characteristics of the circuit overcurrent device must be known in order to assure that this device will promptly clear the ground-fault on a system that is solidly grounded (110.10).

The NEC is not intended as a design specification (90.2(A)). So, the selection of the equipment grounding conductor from 250.118(A), without determining its ability to provide an effective ground-fault current path is not acceptable.

## Isolated Grounding Circuits

The use of isolated grounding circuits as a means of reducing electromagnetic interference (EMI) on the grounding conductor has been referenced in 250.96(B) for many years. At one time, especially for Information Technology Equipment Systems (Article 645), it was considered to be acceptable to provide a separate grounding electrode (system) for this type of equipment. This is not acceptable, as there may be a considerable voltage difference between this separate grounding electrode and the other grounding electrode(s) within the same building or structure. There may be an auxiliary grounding electrode(s) installed as a supplement to the grounding electrode (system) for this type of equipment or to supplement the equipment grounding system, but this auxiliary grounding electrode(s) must be connected to the equipment grounding system supplying this equipment (645.15, 250.54).

The use of isolated grounding circuits (250.146(D)) through the use of isolated ground receptacles may produce the desired results of reducing electromagnetic interference on the equipment grounding conductor. The isolated equipment grounding conductor is permitted to pass through one or more upstream panelboards and terminate at the service equipment or separately derived system within the same building or structure where the circuit(s) originates. The isolated grounding conductor is not permitted to extend beyond the building or structure where it originates. Extending this type of circuit to other buildings or structures may produce significant voltage differences on this grounding system due to the possibility of large voltage differences between buildings or structures due to lightning or other power faults. In addition, the possible increase in the length of the circuit(s) conductors may produce an excessive voltage-drop in the isolated equipment grounding conductor with the resultant excessive voltage-rise on the connected equipment during a ground-fault.

## Permanently Installed Generators

A grounding conductor must provide an effective ground-fault current path (250.4(A)(5), 250.4(B)(4)) from the generator to the first disconnecting means in accordance with 250.35(A) or (B).

Where the generator is a separately derived system, the provisions of 250.30 apply. For a grounded system, an unspliced system bonding jumper connects the grounded conductor to the generator metal frame and its minimum size is determined from Table 250.102(C)(1), (250.30(A)(1)(a)). Or, a Supply-Side Bonding Jumper will be extended with the circuit conductors from the generator to the first disconnecting means enclosure. It's size is based on 250.102(C)(1) (minimum size). In addition, the Supply-Side Bonding Jumper is permitted to be a nonflexible metal raceway or a bus type conductor which has the equivalent cross-sectional area of the wire type conductors specified in 250.102(C)(1). A system bonding jumper will connect the Supply-Side Bonding Jumper to the grounding electrode (system) at the first disconnecting means.

The connection of the grounding electrode conductor (250.66) to the building or structure grounding electrode system is in accordance with 250.30(A)(4), 250.30(A)(5), and 250.8(A).

## Portable, Vehicle Mounted, and Trailer Mounted Generators

The frame of a portable generator is not required to be connected to a grounding electrode where this generator supplies only equipment mounted on the generator and/or cord-and-plug connected equipment through receptacles mounted on the generator (250.34(A)).

This provision is also referenced in OSHA 1926.404(f)(3)(i).

Connecting the metal frame of the portable generator to a grounding electrode may pose a hazard to a person(s) operating a portable tool connected to the generator frame or to a receptacle outlet, as the person(s) may become a parallel path for ground-fault current to return to the source through the earth to this grounding electrode and then through the earth to the generator neutral point.

This provision also applies to vehicle-mounted and trailer-mounted generators (250.34(B)).

Where the portable generator is used to supply fixed wiring systems, the provisions of 250.30 apply. In this case, for a grounded system, there will be the same grounding and bonding requirements from 250.30(A)(1),(2),(3),(4),(5) as for a permanently installed generator and a connection to a grounding electrode is required.

For circuits of over 250 volts to ground (480/277V, 600/346V), the electrical continuity of metal raceways and cable assemblies must be assured in accordance with 250.92(B). This includes wrenchtight threaded connections, threadless couplings and connectors made tight, or the use of bonding-type locknuts, bushings, or through the use of properly sized bonding jumpers.

Where expansion-deflection or deflection fittings are used, properly sized bonding jumpers are normally required to assure electrical continuity (250.98).

The bonding requirements to assure electrical continuity of the normal noncurrent-carrying metal parts of electrical equipment, metal raceways, metallic cable assemblies, and metal enclosures in Hazardous (Classified) Locations is extremely important (250.100, 501.30, 502.30, 503.30, 505.30, and 506.30).

This bonding requirement is important, as even static electricity has more than enough ignition capable energy where flammable gases and vapors are within their lower or upper flammable limit range. And, while it is true that maximum explosion pressures are reached when the mixture of fuel and air are in the mid-range of the lower and upper flammable limits of the gas or vapor, ignition is still a possibility anywhere within the lower and upper flammable limit range.

Section 500.8(E) requires threaded connections of conduits and fittings to be wrenchtight to prevent arcing at conduit joints, which may be a source of ignition.

Section 500.8(E)(1) requires NPT-threaded entries into explosion-proof equipment to be made with at least five threads fully engaged. And the hot ignition gases will flow through the path established by the five fully engaged threads and this allows the enclosure to cool the hot gases before they are released into the outside atmosphere.

This requirement is not only to assure the proper bonding integrity of the threaded entry, but also to create a flame arresting path through the threaded entry.

Explosion-proof equipment is not designed to be air or vapor tight. The enclosure is designed to breathe due to the differences in atmospheric pressure from inside to outside the enclosure, caused by different ambient temperatures. At certain times, there may be ignitable concentrations of fuel/air mixtures within the explosion-proof enclosure. The operation of an arc producing device within this enclosure may be the source of ignition. This enclosure is designed to withstand at least four times the maximum explosion pressure produced within the enclosure, without rupturing or permanent distortion.

The ignition hazard may be increased due to conditions that are uncommon, but certainly always a concern.

One incident that I will always remember occurred in a generating station in 1971. I was working there for an engineering firm at that time. A 100 megawatt natural gas/steam turbine generator exploded one night, causing one death and the destruction of this generator. It was determined that the explosion was caused by a hydrogen leak and the source of ignition was static electricity, as the hydrogen leak caused the disturbance of dust, dirt, and other materials in the vicinity of the leak. The contact and separation of this material caused the static discharge and the hydrogen/air mixture was within the flammable limits of this material. The flammable limits of hydrogen range from 4% to 75% in normal air. However, when the fuel/air mixture is in the mid-range of the lower and upper flammable limits, the explosion pressure may be maximum. Of course, the energy of the ignition source also plays a part in the explosion pressures that are created and static electricity is a low energy source of ignition. In this explosion, the conditions that were present caused complete destruction of this large generator, even though the source of ignition was not a high energy ignition source.

We have made several references to 250.102 when we sized Supply-side and System Bonding Jumpers from Table 250.102(C)(1). Equipment bonding jumpers on the load side of an overcurrent device are sized in accordance with 250.102(D). This Section identifies the bonding jumper size to be in accordance with 250.122 and, where a single common continuous bonding jumper is used to connect two or more raceways or cables, its size is determined in accordance with 250.122, based on the largest size overcurrent device protecting the installed circuits.

However, while the bonding jumper size is not permitted to be smaller than identified in Table 250.122 (250.122(A)), it may be larger, but not larger than the circuit conductors supplying the equipment.

If the ungrounded conductors are increased in size for any reason, except where the ungrounded conductors are increased in size because of increased ambient temperature (310.15(B)) or proximity effects (more than three current carrying conductors in a raceway or cable, 310.15(C)(1)), there must be a proportional increase in the size of the equipment grounding or bonding conductor (250.122(B)).

**Example**

Due to voltage-drop, the ungrounded conductors on a 40 ampere branch-circuit are increased in size from 8AWG, copper to 4AWG, copper. The proportional increase in the size of the wire is a follows:

8AWG, 16,510 cm

4AWG, 41,740 cm

$$\frac{41,740}{16,510} = 2.53$$

Table 250.122-40 ampere circuit-10AWG, copper

$$\begin{array}{r} 10,380\text{ cm} \\ \underline{\times 2.53} \\ 26,261.40\text{ cm} \end{array}$$

The equipment grounding conductor size is 26,261.40 circular mils, 4AWG, 41,740 circular mils (Table 8, Chapter 9).

There is an exception in 250.122(B), that permits the equipment grounding conductor size to be determined by a qualified person, providing that the size of this conductor provides an effective ground-fault current path (250.4(A)(5), 250.4(B)(4)).

Bonding jumpers and equipment bonding jumpers are permitted to be installed inside or outside of a raceway or enclosure (250.102(E)).

However, if installed on the outside of the raceway or enclosure, the length of the bonding jumper is limited to six feet (1.8 m). The magnetic flux density between conductors of an ac circuit decreases with the increased spacing between conductors and the circuit impedance will increase. And this certainly applies to a bonding or an equipment grounding conductor. In fact, it has been determined that where the equipment grounding or bonding conductor is on the outside of the raceway, its impedance will be twice that if it were inside the raceway with the other conductors. This is the reason that 300.3(B) requires all of the conductors of the same circuit, including the grounded conductor, and all equipment grounding and bonding conductors, to be within the same raceway, conduit body, auxiliary gutter, cable tray, wireway, cablebus assembly, trench, cable or cord. By grouping the circuit conductors in this manner there is an increase in the effects of inductive and capacitive coupling between the circuit conductors and a reduction in inductive heating and overall circuit impedance.

Where conductors are installed underground, all of the conductors of the same circuit, including grounded conductors and equipment grounding conductors, are to be installed in the same raceway or cable, or for individual conductors, installed in close proximity in the same trench for important impedance reduction (300.5(I)).

There is a connection here to 250.134, which addresses the grounding of equipment *fastened in place* or connected by permanent wiring methods. Section 250.134(2) states that this equipment is connected to an equipment grounding conductor that is contained within the same raceway, cable, or otherwise run with the circuit conductors.

However, 250.134, Exception No. 1 permits the equipment grounding conductors to be run separately in accordance with 250.130(C). This Section involves the replacement of nongrounding type receptacles or snap switches or branch circuit extensions. In this case, the equipment grounding-type receptacle, a snap switch with an equipment grounding terminal, or a branch circuit extension, may be connected to any accessible point on the grounding electrode system (250.50), or to any accessible point on the grounding electrode conductor, or to an equipment grounding conductor that is part of another branch circuit that originates from the enclosure where the branch circuit for the receptacle, snap switch or branch-circuit originates, or for grounded systems, the grounded service conductor within the service equipment enclosure, and for ungrounded systems, the grounding terminal bar within the service equipment enclosure.

These options sound reasonable on the surface, but they actually are against everything that we have covered so for. We have stressed that on an ac circuit the path for ground-fault current must be in close proximity to the other circuit conductors for important impedance reduction and to facilitate the prompt clearing of the circuit overcurrent device in the event of a ground-fault (300.3(B)) on a solidly grounded system. And, yes, 250.102(E) does permit bonding conductors to be on the outside of raceways or enclosures, where their length does not exceed six feet (1.8 m). So, the use of 250.130(C) for the installation of a grounding type receptacle, snap switch with an equipment grounding terminal, or a branch circuit extension where the equipment grounding conductor is through a path that may be widely separated from the circuit supply conductors, may not provide an effective ground-fault current path and, therefore, should be avoided.

As for as the installation of a bonding conductor or equipment grounding conductor on the outside of a raceway or enclosure, the reasoning behind the 6 foot length restriction is that where this conductor is on the outside of the enclosure, the impedance of the grounding path is twice as much as when this conductor is within the raceway or enclosure with the other circuit conductors. However, due to the limited length of the bonding conductor, the increase in the impedance of the grounding conductor is minimal.

Section 250.134, Exception No. 2 applies to direct current circuits. In this case, only resistance and not inductive reactance (impedance) applies. So, if the equipment grounding conductor is properly sized, this conductor may be run separately from the other circuit conductors. However, as an amendment to 250.134, Exception No. 2 (90.3), 690.43(C) requires the PV system circuit conductors that leave the vicinity of the PV array to comply with 250.134, that is the equipment grounding conductors must be run with the PV circuit conductors.

## Bonding Conductor (Bonding Jumper)

A conductor that ensures the required electrical conductivity between metal parts that are required to be electrically connected. There are many examples where these conductors are used as a ground-fault current path and these bonding conductors must be sized accordingly.

For example, the bonding of the normally noncurrent carrying metal parts that are a part of the service equipment installation. In this case, the metallic raceways, metallic cable assemblies, metallic cable trays, and metallic auxiliary gutters that enclose, contain, or support service conductors (the conductors, overhead or underground, between the service point and the first point of connection to the service-entrance conductors at the building or other structure). For underground service conductors, the point of connection may be inside or outside of the building or structure.

All enclosures containing service conductors are required to be bonded together (250.92(A)).

Where bonding jumpers are installed in accordance with the requirements of Article 250, such as where reducing washers are used for oversized enclosure knockouts or where standard locknuts or bushings are used to connect metallic raceways or cable assemblies to metallic enclosures, a properly sized bonding jumper must ensure the bonding integrity of the service installation.

For example, an installation includes a 1500 kVa, 3-phase transformer. The primary voltage is 13,800 volts and the secondary voltage is 480/277 volts. The transformer nameplate impedance is 3.50% and the worst case impedance is 3.15% (UL 1561).

$$\frac{1500 \times 1000}{831.36(480\text{ V} \times 1.732)} = 1804 \text{ amperes}$$

$$\text{Multiplier} = \frac{100}{3.50 \times 0.90} = 31.746$$

$$\begin{array}{rl} 1804 & \\ \times 31.746 & \text{amperes} \\ \hline 57,270 & \text{amperes} \end{array}$$

There is a significant motor load served in this example, so the full-load current is increased by 4 to accommodate the rotational energy of the motor load.

$$\begin{array}{rl} 1804 & \\ \times 4 & \text{amperes} \\ \hline 7216 & \text{amperes} \end{array}$$

$$\begin{array}{rl} 57,270 & \\ +\ 7216 & \text{amperes} \\ \hline 64,486 & \text{amperes} \end{array}$$

This is the available fault current at the transformer secondary.

There are 6-500 kcmil copper conductors per phase, in Schedule 80-PVC conduits, extending 100 feet (30 m) to the service disconnect (2000 ampere, fusible switch, UL Class L-KRP-C fuses, interrupting rating – 300,000 amperes (110.9).

$$\frac{1.732 \times 100 \text{ feet} \times 64,486 \text{ amperes}}{26,706 \text{ ("C" value)} \times 6 \text{ (per phase)} \times 480 \text{ V}} = \frac{11,168,975}{76,913,280} = 0.1452$$

$$\text{Multiplier} = \frac{1}{1 + 0.1452} = 0.8732$$

$$\begin{array}{rl} 64,486 & \text{amperes} \\ \times 0.8732 & \\ \hline 56,309.18 & \text{amperes (at service disconnect)} \end{array}$$

The service equipment must be legibly marked in the field with this available fault current (110.24(A)), and this is the required short-circuit current rating of the service equipment.

The interrupting rating of the service fuses must be at least 56,309.18 amperes and the service fuses exceed this requirement, as their interrupting rating is 300,000 amperes (110.9).

The equipment bonding jumper for this installation must be sized to withstand no less than 56,309.18 amperes. The service conductors are installed in nonmetallic raceways, so there will be no bonding locknuts or bonding bushings. There will be bushings on the PVC conduits within the service equipment (352.46). However, there will be connections to a terminal bar within the service equipment and this terminal bar must be able to carry 56,309.18 amperes without damage (110.10). The service overcurrent device (fuses) are current-limiting and will likely interrupt this fault current in less than half a cycle (0.008 seconds).

The bonding of equipment to the grounded service conductor is in accordance with 250.8(A). There are eight acceptable connection methods referenced here, including the exothermic welding process that we have discussed earlier. Machine screw-type fasteners that engage not less than two threads or are secured with a nut are acceptable.

Ground clamps and fittings that are exposed to physical damage are required to be protected by being enclosed in wood or metal, or an equivalent protective covering (250.10, 250.53(A)(4)).

The bonding of threaded metal raceways that contain service conductors must be assured through the use of connections that are made wrenchtight. This includes threaded couplings and listed threaded hubs on enclosures (UL 467).

Threadless couplings and connectors that are made up tight for metal raceways and metal-clad cables (UL 467) are acceptable.

Bonding type locknuts and bonding type bushings are acceptable with properly sized bonding jumpers (250.92(B),(2),(3),(4)).

## Article 517: Health Care Facilities

Section 517.18(B)(1) requires that each patient bed location be provided with a minimum of eight receptacles. These receptacles may be single, duplex, or quadruplex type, or any combination of these types and these receptacles must be listed as "hospital grade" (517.18(B)(2)). Each patient bed location must have no less than two branch circuits, one from the critical branch and one from the normal branch.

Even though 250.118(A)(1) permits an equipment grounding conductor to be insulated, covered, or bare, 517.13(B)(1) requires the equipment grounding conductor for the receptacles in patient care spaces to be insulated and identified along its entire length by green insulation.

In addition, 517.13(A) requires that the branch circuits serving patient care spaces be provided with a wiring method that provides an effective ground-fault current path (250.4(A)(5)). These branch circuit conductors must be installed in a metal raceway or metal cable, that in itself, qualifies as an equipment grounding conductor from 250.118. However, this does not imply that the metal raceway or metal cable provides an effective ground-fault current path. This must be determined by calculation and not just because these wiring methods are referenced in 250.118(A).

In addition, equipment bonding jumpers that are used in patient care spaces must be insulated and sized in accordance with 250.122. Section 517.14 requires the equipment ground terminal bus bars of the normal and essential branch circuit panelboards to be bonded with an *insulated* continuous copper conductor not smaller than 10AWG. This bonding conductor must have the required current-carrying capacity to *safely* carry the fault current that may be imposed.

The essential electrical system (Type 1) is required to have two independent sources of power (517.30) and, in most cases, the alternate power source is a generator (517.30(A)). Utility power is the normal source and a generator is the alternate source.

The normal source will be connected to a grounding electrode system and it is very important that the alternate power source be connected to the same grounding electrode system. This will assure that the connected electrical load has the same ground (earth) reference and there will not be a ground potential difference between these separate supply systems. So, any bonding conductor connections from the normal source and the alternate source to the grounding electrode system must be through conductors that are as short and straight as practicable.

## DC System Grounding and Bonding

In a manner that is similar to ac systems, 250.162 requires the grounding of dc systems as follows:

Section 250.162(A): Two-wire dc systems operating at greater than 60 volts but not greater than 300 volts are required to be grounded. This does not apply to industrial equipment in limited areas that is equipped with a ground detector and this equipment is installed immediately adjacent to, or integral to, the supply sources (250.162(A), Exception No. 1).

A rectifier-derived dc system in compliance with 250.20 is not required to be grounded. For example, where the rectifier is supplied by an ac system and the maximum voltage-to-ground exceeds 150 volts and is not required to be grounded in accordance with 250.20(B)(1).

Direct-current fire alarm circuits, where the maximum current does not exceed 0.030 ampere, do not require grounding (Part III, Article 760).

Three-wire direct-current systems are required to be grounded where they supply premises wiring systems. Keep in mind that premises wiring systems include interior or exterior wiring, including power, lighting, control and signal circuit wiring, together with all associated hardware, fittings, and wiring devices both permanently and temporarily installed. Generally, this wiring is supplied from the service point (the point of connection between the facilities of the serving utility and the premises wiring). However, the premises wiring may also be supplied from other sources, such as portable generators, stand-alone generators, and solar photovoltaic systems.

If the dc system is an on-premises source, the grounding connection is required to be made at the source, or at the first system disconnect or overcurrent device, or the grounding connection may be made by other means through equipment that is listed and identified for this purpose (250.164(B)).

The size of the grounding electrode conductor is in accordance with 250.166, which is similar to 250.66. However, the grounding electrode conductor is required to be not smaller than the dc neutral conductor and not smaller than 8AWG copper (8.37 mm$^2$) or 6AWG aluminum or copper-clad aluminum (13.30 mm$^2$) (250.166(B)).

Where the dc system is grounded by connection to a ground rod, pipe, or plate and this is the sole connection to these grounding electrodes, the grounding electrode conductor, may be 6AWG copper (13.30 mm$^2$) or 4AWG aluminum or copper-clad aluminum (21.15 mm$^2$) (250.166(C)). Also, where aluminum or copper-clad aluminum conductors are used, these conductors are not normally permitted within 18 inches (450 mm) of the earth where these conductors are external to buildings (250.64(A)(3)).

Where the grounding electrode is a concrete-encased electrode, the conductor that is the sole connection to this electrode will be 4AWG copper (21.15 mm$^2$) (250.166(D)).

In addition, where the grounding electrode is a ground ring, the conductor that is the sole connection to this electrode will be at least the same size as the conductor used for the ground ring, and not smaller than 2AWG bare copper (33.63 mm$^2$) (250.166(D)).

Ungrounded dc systems are required to have ground-fault detector systems (250.167(A)).

Grounded dc systems are permitted to have ground-fault detection (250.167(B)).

A direct-current system bonding jumper must be used to connect the equipment grounding conductor to the grounded conductor. This is an unspliced bonding jumper and this connection may be provided at the source or at the first disconnecting means where the dc system is connected to ground (250.168).

For dc systems that are ungrounded and serve as separately derived systems from a stand-alone power source, a grounding electrode conductor is required to connect the metal enclosures, raceways, cables, and exposed non-current carrying metal parts of equipment to a grounding electrode (system). The grounding electrode conductor size is in accordance with 250.166. It should be noted that this provision does not apply for portable and vehicle-mounted generators in accordance with 250.34(A)(B). That is where the frame of these generators is not required to be connected to a grounding electrode where the generator supplies only equipment mounted on the generator or cord-and-plug connected equipment through receptacles mounted on the generator, or both (250.169, OSHA 1926.404(f)(3)(i)).

## AC Substation Grounding and Bonding

IEEE Standard 80 covers the very important topic of AC substation grounding and bonding. Because of the medium and high voltages that are present in these electrical installations, proper grounding and bonding is essential to protect the people working in and around a substation or switchyard and, also, to protect the connected equipment.

I will begin this study with my experience in the installation of a 230 kV switchyard. The installation included two autotransformers, multiple circuit switchers, connections to 8–80 megawatt gas turbine driven generators, extensive overhead aluminum bus, as well as smaller transformers for 13.8 kV to 480 V distribution systems.

The grounding and bonding system was designed by the utility and in accordance with IEEE Standard 80. However, there were additional safety measures added in this design.

The composition of the soil was primarily cinder fill. This material is conductive and a soil resistivity test indicated a resistivity of approximately 9000 ohm-centimeters (the resistivity across the faces of a cubic centimeter of this soil). Unfortunately, this type of soil is also highly corrosive. So the use of tinned copper conductors was a part of the design.

The soil resistivity is determined by the use of an earth (ground) resistance tester.

Four, one foot (30.48 cm) long ground rods are driven into the earth and spaced 20 feet (609.6 cm) apart in a straight line.

The ground resistance tester has four terminals, marked C1, P1, P2, and C2, and the test leads are connected from these terminals to the ground rods.

The resistance is measured and the soil resistivity is calculated by the formula:

$Rho = 2\pi AR$

$A =$ the distance between the electrodes in centimeters

$R$ = the resistance recorded by the meter.

For example, where the resistance reading is 4 ohms, the soil resistivity will be as follows:

$$
\begin{array}{rl}
6.2832 & (2 \times 3.1416) \\
\times 609.60 & \text{cm (20 feet} \times \text{12 inches} \times \text{2.54 cm)} \\
\times \dfrac{4}{15,321} & \begin{array}{l}\text{(ohms)} \\ \text{ohm centimeters}\end{array}
\end{array}
$$

A $\frac{5"}{8}$ (1.5875 cm) diameter 10 foot (3.048 m) long copper-clad steel ground rod driven into this soil will have an approximate resistance-to-ground of 50.73 ohms (250.53(A),(1),(2),(3),(4)).

The grounding and bonding grid consisted of 4/0 (107.22 mm$^2$) bare tinned copper conductors on 10 foot (3.0 m) centers arranged in a grid pattern with the joints exothermically welded (110.14). The grid was buried to a depth of three feet (1.0 m), with the soil topped with a layer of large stones to afford additional protection from step and touch potential differences. The people walking and working in this environment will have the bottom of their footwear only making contact with the sharp projections of the large stones, so there is minimal contact with the surface. The grounding and bonding grid was supplemented with 40 foot long (12.20 m) creosoted wood pilings, with every third piling supplemented with a 4/0AWG tinned copper conductor, which was connected to the ground grid with an exothermic welded connection.

The switch-gear was bonded on opposite corners to the ground grid with 4/0AWG copper conductor connections. This was also done with the auto-transformers, step-down transformers, other large equipment frames, and the steel columns which served as the support for the insulators and the over-head aluminum bus. These bonding conductor connections were installed short and straight, avoiding unnecessary and sharp bends which would increase the impedance of these connections and affect the voltage-rise above earth potential during faults and lightning induced currents.

The grounding and bonding of the metal fence surrounding this installation is always a concern. In addition, in order to protect people from the effects of touch and transferred potential differences, the grounding and bonding grid was extended 3 feet (1.0 m) beyond the metal fence. The metal fence was installed within 16 feet (5.0 m) from the equipment within the switchyard, so the bonding of the metal fence to the grounding grid is a requirement (IEEE 80, 250.194(A)). The fence fabric, when properly bonded to the grounding and bonding grid, actually adds to the cross-sectional area of this grid and serves to lower the resistance-to-ground of the entire system.

Section 250.194(A)(1) requires a bonding connection from the metal fence to the grounding electrode system at each fence corner and at maximum 160 foot (50 m) intervals along the fence. Where overhead conductors cross the fence, there must be bonding jumpers on each side of the crossing (250.194(A)(2)). In our installation, the bonding from the metal fence to the bonding grid was done at 50 foot (15 m) intervals.

The bonding of the metal fence support posts at the gate entrances to the grounding electrode systems, and the proper bonding of the fence gates to their support posts is also a requirement (250.194(A)(3)). There is also a requirement to install a buried bonding jumper across the opening of a gate or other opening in the fence (250.194(A)(4)).

In the example of our switchyard, the grounding and bonding grid extended beyond the swing of the fence gates (250.194(A)(5)) and there was no barbed wire above the fabric of the fence in our switchyard. If barbed wire had been installed, a bonding connection to the grounding electrode system is required (250.194(A)(6)).

Erico (Pentair Plc) has the proper grounding and bonding connections for this substation, switchyard environment. Finally, if there is a possibility for potential differences to be transferred through the metal fence to other areas, it will be necessary to interrupt the continuity of the metal fence, possibly with a section of wood fencing. Grounding conductor connections must be listed, labeled, or both and used in accordance with any instructions that are included and associated with the listing or labeling (110.3(A),(B),(C)).

## Information Technology Equipment

Section 645.14 is entitled System Grounding for listed Information Technology Equipment, which is in accordance with Parts I and II of Article 250. For systems like these, the bonding and grounding requirements of 250.30 for separately derived systems will normally apply. However, 645.14 states that, “power systems derived from listed information technology equipment that supply information technology systems through receptacles or cable assemblies supplied as part of this equipment, shall not be considered *separately derived* for the purpose of applying 250.30”. This does not mean that the bonding and grounding requirements of these systems is not safely done. The key word of 645.14 is “listed” and when the equipment is listed, labeled, or both, or identified for a specific use and used in accordance with the listing instructions, the provisions of 110.3(B) are satisfied.

Section 645.15 covers the equipment grounding and bonding of information technology equipment. The first sentence of this section states that all exposed noncurrent-carrying metal parts of an information technology system shall be bonded to the equipment grounding conductor or shall be double insulated. Here is where there is a reference to a signal reference structure (grid). Where signal reference structures are installed, they are required to be bonded to the equipment grounding conductor (system) for the information technology equipment. This is the "single-point ground" for the information technology equipment. It is extremely important that the connections to the signal reference structure be accomplished in a specific method.

Signal reference structures have been used for quite some time. Initially, they were constructed with bare copper conductors and arranged in a grid pattern on the concrete sub-floor beneath the raised floor of the information technology equipment room. The size of the SRG conductors were commonly 4AWG (21.15 $mm^2$) copper and placed on approximately 4 foot (1.2 m) centers. This bonding grid was bonded to the metallic floor pedestals as well as to the metal frames of the equipment grounding system for this equipment. On the surface, this appears to be an effective method of providing an equipotential plane for this type of installation. However, the inherent problem with this type of installation was not with the size of the wire (4AWG copper), but the fact that this wire was round. For example, a 4/0AWG copper conductor (107.22 $mm^2$) has a dc resistance of 0.0608 ohms (1000 feet) (305 m) at 75 °C (Table 8, Chapter 9) and an ac impedance of 0.243 ohms (ohms-to-neutral) at 0.85 power factor where the conductors are installed in a PVC conduit (Table 9, Chapter 9). However, at 10 mHz, and a length of only 10 feet (3.05 m), the ac impedance of this conductor is 232 ohms. This short length of conductor cannot equalize voltages across the ends of the cable. The "skin effect" of this current flow at this high frequency is the problem.

In order to make the bonding connections more effective, these conductors must expose as much of their cross-sectional area to their surface as possible. Therefore, the signal reference grid must be made flat and not round.

The flat signal reference grid consists of flat copper conductors on two to four foot centers and installed directly on the concrete sub-floor. Every sixth floor pedestal is bonded to the signal reference grid with short and flat conductors. The signal reference grid can be constructed in 16 foot widths, with any additional sections (where necessary) exothermically welded in order to completely cover the concrete sub-floor. In some information technology equipment rooms, this SRG is also installed in the walls of this room (Faraday Cage). The metal raceways, cable trays, and other metallic piping systems are bonded to the signal reference grid. The key to all of these bonding connections, in addition to being flat, is that they are no longer than necessary to complete the bonding connections.

An added benefit of the use of the flat conductor signal reference grid is that this gird is magnetically and capacitively coupled to the conductors, metal raceways, and cable assemblies that lay on top of the signal reference grid, providing a low impedance connection to the signal-point grounding system within the information technology equipment room. In addition, any current flow, regardless of the frequency, through the rebar in the concrete sub-floor will be magnetically and capacitively coupled into the flat signal reference grid on top of the concrete sub-floor and directed to the single-point ground.

The flat signal reference grid is bonded to the single point grounding system within the information technology equipment room with a short flat conductor of the same size as the signal reference grid.

## Agricultural Buildings, Equipotential Bonding

These types of buildings or structures are very often supplied by a feeder or branch-circuit(s). Section 250.32(A) will apply in this case and a grounding electrode (system) will be installed in compliance with 250.50 and 250.52. There is an exception that waives this requirement if only a single branch-circuit, including a multiwire branch circuit, supplies the building or structure and this branch-circuit includes a properly sized equipment grounding conductor for the purpose of grounding the metal parts of equipment.

An equipment grounding conductor is required to be included and run with the building or structure supply conductors and its minimum size is in accordance with 250.122. And this conductor will be any of the types referenced in 250.118. However, just as in the other examples in this book, the equipment grounding conductor must provide an effective ground-fault current path (250.4(A)(5)), (250.32(B)(1)).

These types of installations normally require an equipotential plane(s), due to the sensitivity of livestock to even slight voltage differences, especially where concrete floors are damp or wet and the livestock may be in contact with metallic equipment that may become energized.

Wire mesh embedded in the concrete floor or bonding conductors embedded in the concrete and arranged in a mesh pattern may serve this purpose.

The bonding conductor connection is required to be at least 8AWG (solid) and this conductor may be insulated, covered, or bare. The bonding conductor connections must be listed (110.3(B),(C)) for this purpose and in compliance with 110.14 and 250.8. There is no requirement and no need to connect this bonding conductor to a grounding electrode. The sole purpose of this bonding system is to limit the effects of voltage differences and to provide protection for the livestock. Connecting this bonding conductor to a grounding electrode will not improve the means of limiting voltage differences. This purpose is much the same as the equipotential bonding system for a swimming pool, that is to equalize the voltage gradient in the area that is accessible to livestock (547.44(B)). So, the equipotential plane is grounded, but not to a separate and dedicated grounding electrode.

The equipotential plane(s) is required to be bonded to the grounding electrode system or, where the building is served by a panelboard, the bonding conductor connection may be made to the equipment grounding system of the panelboard (547.44(B)).

## Swimming Pools, Equipotential Bonding

Section 680.26(A) covers the equipotential bonding requirements for swimming pools, the purpose of which is to limit voltage gradients (differences) in the pool area.

The bonded parts include (1) conductive pool shells, where the pool shell is conductive, such as unencapsulated reinforcing steel, and bonded by steel tie wires to form the common bonding grid.

A copper conductor grid constructed of 8AWG bare solid copper conductors bonded at all crossing points with fittings that are listed and indentified for this purpose in accordance with 250.8 can be the common bonding grid. These conductors are arranged to conform to the shape of the pool and arranged in a 12 inch (300 mm) by 12 inch (300 mm) grid pattern.

The perimeter surface extends 3 feet (1 m) horizontally from the inside walls of the pool. This includes the reinforcing rods or steel mesh within the perimeter concrete surface. The bonding conductor connection is through a 8AWG solid copper (or larger) conductor with the connections in compliance with 250.8. Section 680.26(B)(6) identifies the electrical equipment that is required to be bonded.

1. Electrically powered pool covers
2. Pool water recirculating equipment
3. Electrical equipment within 5 feet (1.5 m) measured horizontally from the inside wall of the pool, or 12 feet (3.7 m) measured vertically above the maximum water level of the pool.

Once again, the purpose of these bonding requirements is to reduce the effects of voltage differences in the area of the pool. Of course, any electrical equipment associated with the pool will be properly grounded. So, there is no need to connect the bonding system conductors to a grounding electrode (system).

In addition, 250.52(B)(3) prohibits the use of the swimming pool bonding system as a grounding electrode, as any voltage-rise on the pool bonding system may prove to be hazardous to people using the pool.

CHAPTER

# 2

## Questions: Bonding Systems and Equipment

1. Which of the following elements addressed by the NFPA 780 Standard for the Installation of Lightning Protection Systems, are referenced in 250.4(A)(1)?

   (a) Grounding
   (b) Bonding
   (c) Both a and b
   (d) None of the above

2. The rebar of a concrete-encased electrode *may* be used as a conductor to interconnect electrodes of a grounding electrode system.

   (a) True
   (b) False

3. A grounding electrode is required at a separate building or structure when it is served by a multi-wire branch circuit.

   (a) True
   (b) False

4. The bonding jumper(s) that are installed to interconnect the grounding electrode system have a minimum size in accordance with _______________ and are installed without a joint or splice.

   (a) 250.53(C)
   (b) 250.64(E)
   (c) 250.66
   (d) 250.92(A)

5. Which of the following shall not be used as a grounding electrode?

   (a) Metal underground gas piping systems
   (b) Aluminum

(c) Metal well casings
(d) Both a and b

6. Grounding electrode conductors and grounding electrode bonding jumpers that are in contact with _______________ are not required to comply with 300.5, but must be buried or protected if they are at risk of physical damage.

(a) Water
(b) The earth
(c) Metal
(d) Concrete

7. Ferrous metal raceways and enclosures for grounding electrode conductors must be bonded at both ends to the grounding electrode or grounding electrode conductor to establish a(n) _______________ parallel path.

(a) Mechanically
(b) Electrically
(c) Physically
(d) None of the above

8. If the grounding electrode conductor or bonding jumper connected to one or more rod, pipe, or plate electrodes does not extend to other types of electrodes that require a larger conductor size, the grounding electrode conductor is not required to exceed a _______________ AWG copper wire.

(a) 10
(b) 8
(c) 6
(d) 4

9. The metal structural frame of a building is allowed to be used as a conductor to interconnect electrodes within the grounding electrode system, or as a grounding electrode conductor. Hold-down bolts that secure the structural steel columns and connect to a concrete-encased electrode (as per 250.52(A)(3)) located in the support footing or foundation are permitted to link the building's metal structural frame to the concrete-encased grounding electrode.

(a) True
(b) False

10. An external _______________ means for connecting intersystem conductors must be provided at the service equipment or metering equipment enclosure, as well as the disconnecting means for buildings or structures supplied by a feeder.

(a) Bonding
(b) Ungrounded
(c) Secondary
(d) Both a and b

11. Metal water piping systems must be bonded to the ______________, or to one or more grounding electrodes used, providing the grounding electrode conductor or bonding jumper to the grounding electrode is of adequate size.

 (a) Grounded conductor at the service
 (b) Service equipment enclosure
 (c) Grounding electrode conductor, if sufficiently sized
 (d) Any of these

12. Bonding devices used for the grounding of the metal frames of PV modules and other equipment must be ______________

 (a) Listed
 (b) Labeled
 (c) Identified
 (d) All of the above

13. According to NFPA 70E-2024-Annex 0, impedance grounding is effective in reducing:

 (a) Ground-faults
 (b) Arc-flash hazards
 (c) Overcurrent
 (d) Overvoltages

14. Normally, there is no overcurrent protection on the supply-side of the service conductors and these conductors have only ______________ protection provided by the transformer primary overcurrent protection.

 (a) Short-circuit
 (b) Overload
 (c) Ground-fault
 (d) Overcurrent

15. A grounding electrode for strike termination devices (lightning protection) must be not less than ______________ feet from any other electrode of another grounding system.

 (a) 5 feet
 (b) 10 feet
 (c) 6 feet
 (d) 8 feet

16. Where a separate grounding electrode is provided for radio and TV equipment (Article 810), what is the minimum size bonding jumper required for connection to the building grounding electrode system?

 (a) 6AWG
 (b) 8AWG
 (c) 12AWG
 (d) 10AWG

17. Bonding devices used for the grounding of the metal frames of PV modules and other equipment must be ______________.

    (a) Listed
    (b) Labeled
    (c) Identified
    (d) All of the above

18. A grounding electrode for the service is not required to be bonded to a grounding electrode for the communications system if this electrode is more than 50 feet away.

    (a) True
    (b) False

19. The metal forming shell (structural reinforcing steel) of a swimming pool may be used as part of the grounding electrode system.

    (a) True
    (b) False

20. The frame of a portable generator ______________ be required to be connected to a grounding electrode where the generator supplies only equipment mounted on the generator, or cord-and-plug connected equipment supplied from receptacles mounted on the generator.

    (a) Shall
    (b) Shall not
    (c) May
    (d) May not

21. A system bonding jumper is associated with ______________

    (a) A service supplied system
    (b) A separately derived system
    (c) Table 250.102(C)(1)
    (d) b and c

22. If rock bottom is encountered and driving a ground rod at up to 45 degrees is not possible, the rod may be buried horizontally in a trench that is ______________ deep.

    (a) 1 foot
    (b) 2 feet
    (c) 2.5 feet
    (d) 3 feet

23. Multiple ground rods should be spaced at least twice the driven length of the longest rod for optimum paralleling efficiency.

    (a) True
    (b) False

24. It is a requirement to extend the solid 8AWG copper bonding conductor for equipotential bonding of a pool or spa to the service equipment.

    (a) True
    (b) False

25. _____________ fully-engaged threads create a flame-arresting path and assures proper bonding for conduit connections in a Class I Division 1 Hazardous (Classified) Location for NPT-threaded entries.

    (a) 5
    (b) 4.5
    (c) 4
    (d) 3.5

26. A nonferrous metal raceway may be used for physical protection of a grounding electrode conductor without bonding at each end.

    (a) True
    (b) False

27. A copper main bonding jumper for a service supply consisting of 350 kcmil copper conductors shall be a minimum size of _ .

    (a) 2AWG
    (b) 2/0AWG
    (c) 4AWG
    (d) 3/0AWG

28. When replacing a non-grounding type receptacle where there is no equipment grounding conductor available in the box, the permitted options include:

    (a) Non-grounding type receptacle
    (b) Grounding-type receptacle
    (c) GFCI-type receptacle
    (d) All of the above

29. The structural metal frame of a building or structure may be used as an equipment grounding conductor.

    (a) True
    (b) False

30. If an auxiliary grounding electrode is used as a supplement to an equipment grounding conductor specified in 250.118, the _ requirements of 250.50, 250.53(C), and 250.58 do not apply.

    (a) Grounding
    (b) Bonding
    (c) Supplemental
    (d) All of these

## Answer Key

1. a (250.4(A)(1))
2. b (250.53(C))
3. b (250.32(A), Exception)
4. c (250.66)
5. d (250.52(B)(1),(2))
6. b (250.64(B)(4))
7. b (250.64(E)(1))
8. c (250.66(A))
9. a (250.68(C)(2))
10. a (250.94(A))
11. d (250.104(A)(1))
12. d (690.43(A))
13. b (250.20(B), Informational Note; 250.36, Informational Note)
14. b (230.90)
15. c (250.53(B))
16. a (810.21(J))
17. d (690.43(B))
18. b (250.50, 250.58, 800.100)
19. b (250.52(B)(3), 680.26(B)(1),(2))
20. b (250.34(A))
21. d (250.30(A)(1))
22. c (250.53(A)(4))
23. a (250.53(A)(3))
24. b (680.26(B))
25. a (500.8(E)(1))
26. a (250.64(E)(1))
27. a (Table 250.102(C)(1), 250.28)
28. d (406.4(D),(1),(2),(3); 250.130(C))
29. b (250.118(B)(2))
30. b (250.54)

# Index